Steverson

MW01536897

Stevenson

A FIRST LOOK AT BIRD NESTS

By Millicent E. Selsam
and Joyce Hunt

ILLUSTRATED BY HARRIETT SPRINGER

WALKER AND COMPANY ✷ NEW YORK

First published in the United States of America
in 1984 by the Walker Publishing Company, Inc.

Published simultaneously in Canada by John Wiley & Sons
Canada, Limited, Rexdale, Ontario.

Library of Congress Cataloging in Publication Data

Selsam, Millicent Ellis, 1912–
    A first look at birds' nests.

    (A First look at series)
    Includes index.
    Summary: An introduction to the many places birds
make their nests, such as chimneys, cliffs, bushes,
traffic lights, and window ledges, and the unusual
things that might be built into the nests.
    1. Birds—Nests—Juvenile literature. [1. Birds—
Nests] I. Hunt, Joyce. II. Springer, Harriett, ill.
III. Title. IV. Series: Selsam, Millicent Ellis,
1912–   . First look at series.
QL675.S38   1984   598.2'56   84-15238
ISBN 0-8027-6565-3

Printed in the United States of America

10 9 8 7 6 5 4 3 2 1

A *FIRST LOOK AT* SERIES

Each of the nature books in this series is planned to develop the child's powers of observation—to train him or her to notice distinguishing characteristics. A leaf is a leaf. A bird is a bird. An insect is an insect. That is true. But what makes an oak leaf different from a maple leaf? Why is a hawk different from an eagle, or a beetle different from a bug?

Classification is a painstaking science. These books give a child the essence of the search for differences that is the basis for scientific classification.

For Daniel Howard Selsam

The authors wish to thank Mr. Stuart Keith, Department of Ornithology, American Museum of Natural History, for reading the text of this book and offering many helpful suggestions.

Birds lay eggs and feed their young in *nests*.

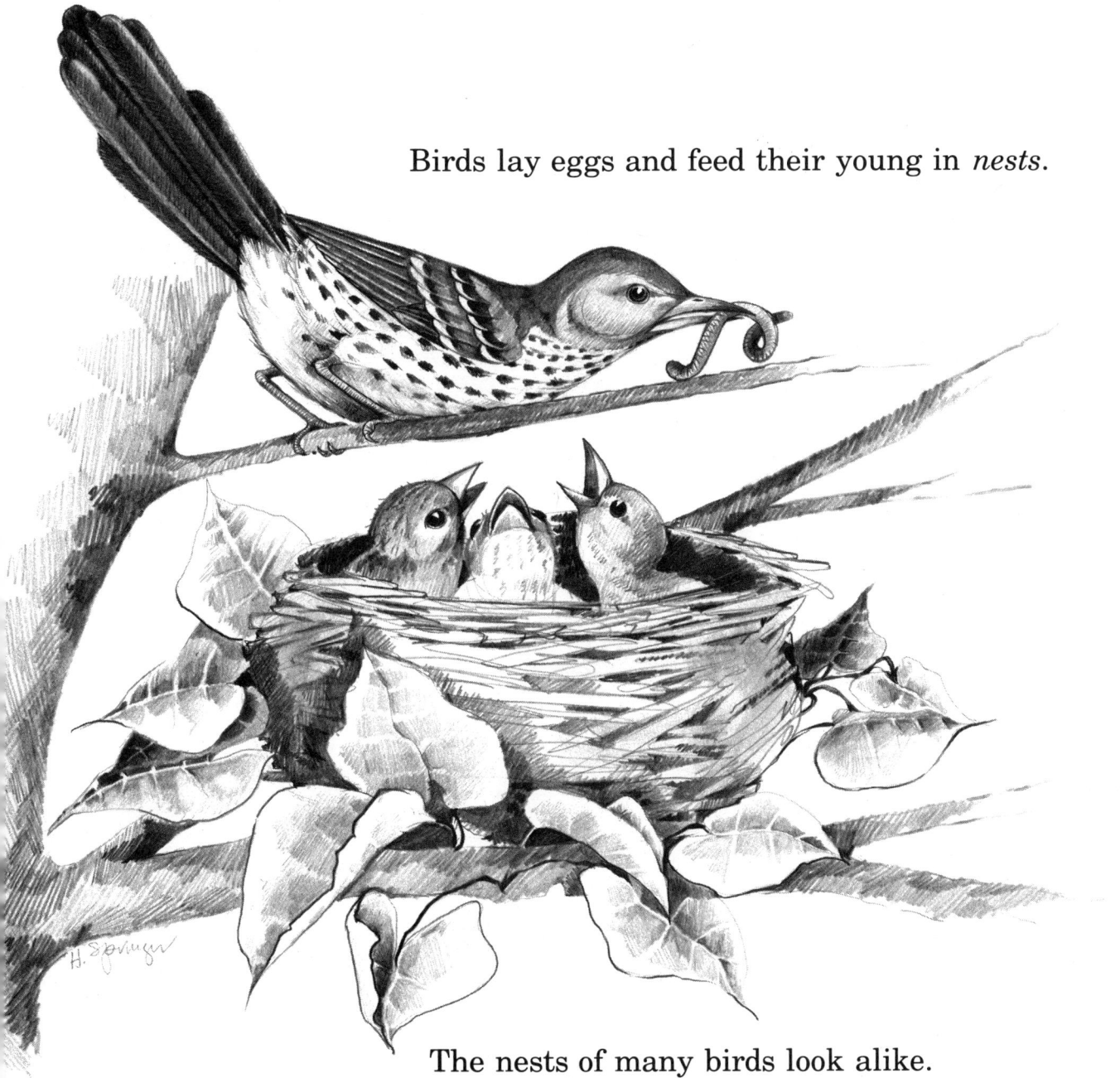

The nests of many birds look alike.
How do you tell them apart?
There are clues to look for.

One important clue is *where* the nest is found.

Would a gull make a nest in a hole in a tree?

No. A gull eats fish, clams and crabs.
It makes its nest close to the seashore.

Would a woodpecker make a nest
on the sand near the sea?

No. A woodpecker eats insects that live in trees.
It makes its nest in tree holes.

*Most birds make their nests close to
where they find their food.*

Many birds make their nests in woodlands.
Woodlands have different layers like the
floors of a building.

Some warblers make their nests
on the ground floor.

The nests of cardinals are above them.

The nests of crows are still higher.

CROW

CARDINAL

WARBLER

Here is a hole in a tree.
Many birds nest in such a hole.
It is hard to tell who is inside
but you can watch the hole to see who goes in.

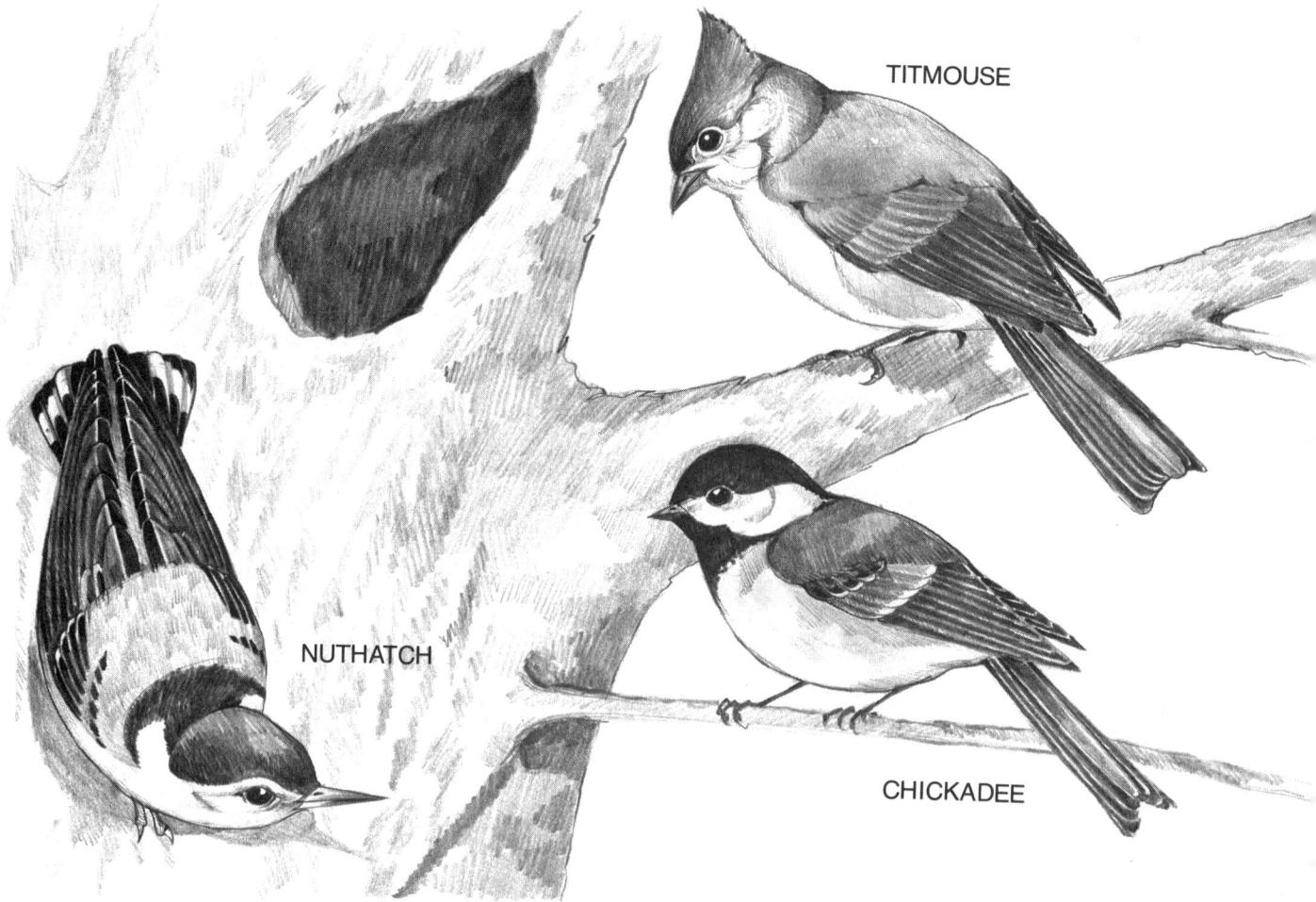

TITMOUSE

NUTHATCH

CHICKADEE

Here is another hole. But this time it is in
a giant cactus.
A woodpecker made it and then moved out.
An elf owl moved in.

Some birds make their nests in meadows
where there are hardly any trees but lots of grass.

The nest of the meadowlark has a roof.
The nest of the bobolink does not.
Which is which?

The nest of the bobwhite is much smaller
than the nest of the pheasant.
Which is which?

Many birds make their nests close to water.
Find the swan's nest among the reeds.
Find the tern's nest on the beach.
Find the grebe's (GREEBS) nest floating on the lake.

White pelicans and flamingos also nest close to water.
Hundreds of these birds nest together in *colonies*.

The nests of the white pelicans look like piles of sticks.

The nests of the flamingos look like pots made of mud.

Although birds usually make their nests
in the country or along the seashore,
their nests can be found in cities too.

Look for the pigeon's nest on the window ledge.

Look for the robin's nest in the bushes.

Look for the bluejay's nest in the park tree.

Look for the house sparrow's nest in the traffic light.

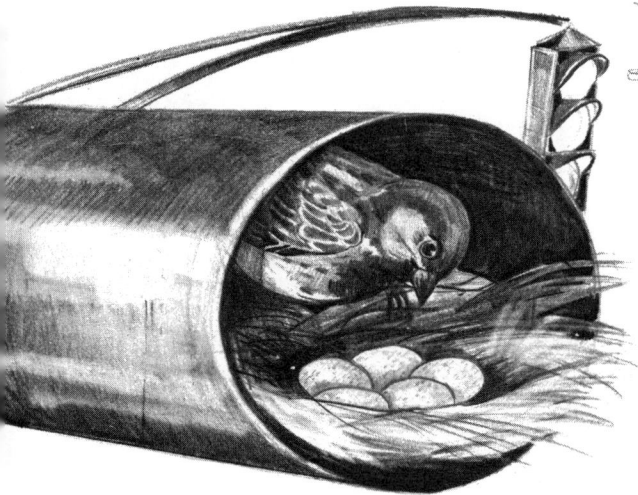

19

The names of some birds can tell you
where their nests are.

CHIMNEY SWIFT

CLIFF SWALLOW

BARN SWALLOW

Chimney swifts make nests in chimneys.
Barn swallows make nests in barns.
Cliff swallows make nests in cliffs.

Match the bird to its nest.

Nests are usually made with grass, twigs, roots and bark that birds gather from the plants around them.

But some birds put strange things in their nests.

Find the snakeskin.
Find the spider web.
Find the horsehair.
Find the paper.

WOODTHRUSH NEST

GREAT-CRESTED FLYCATCHER NEST

VIREO NEST

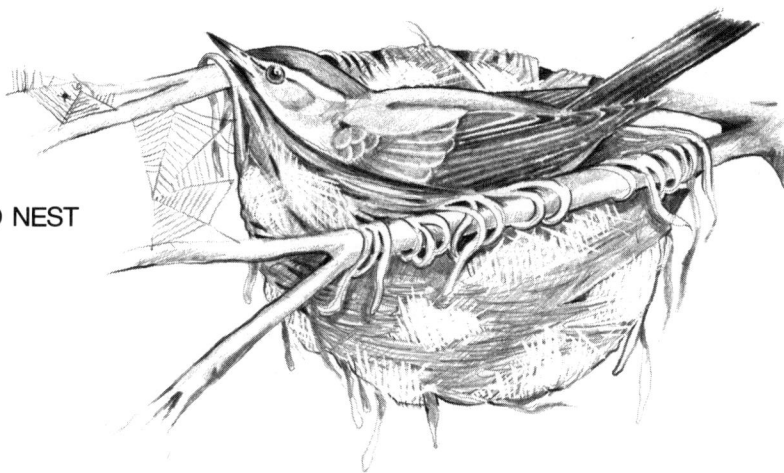

CHIPPING SPARROW NEST

STARLING NEST

In the city, plastic, wire, hairpins, rags and strings
find their way into nests as well.
How many strange things can you find?

The size of the nest is another clue.
The bald eagle's nest is so large that you and lots
of your friends can make a circle around it
if it is laid on the ground.

The hummingbird nest is as
small as half a Ping-Pong ball.
(Its eggs are the size of jellybeans.)

The shape of a bird nest can help you
tell one from another.

Which nest looks like a cup?
Which nest looks like a big round ball?
Which nest looks like a lacy bag?
Which nest looks like a tunnel?

MAGPIE NEST

KINGFISHER NEST

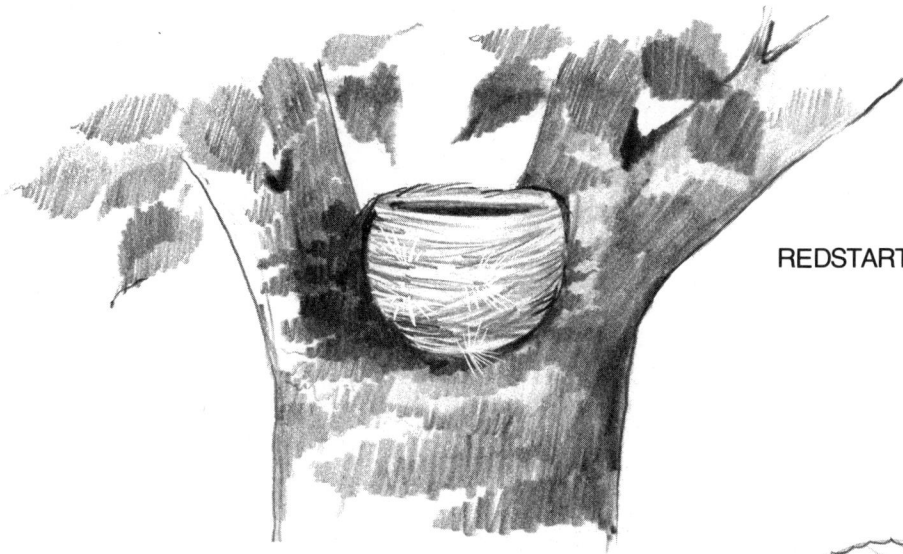

REDSTART NEST

ORIOLE NEST

27

There are birds that make hardly any nest at all.

The killdeer puts a few pebbles on the ground.

The mourning dove puts a few sticks on the branch of a tree.

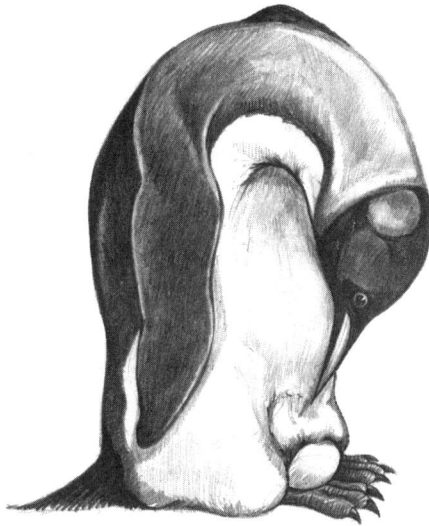

The emperor penguin keeps its one egg on its feet.

Many birds lay their eggs in birdhouses made by people.

PURPLE MARTIN

SCREECH OWL

HOUSE WREN

FLICKER

To tell bird nests apart: See where the nests are found.

They could be

in meadows,

in woodlands,

along the shore,

in cities.

Look at the size of the nests.

Look at the shape of the nests.

See what the nests are made of.

Look for colonies of nests.

BIRD NESTS IN THIS BOOK